GERMAN TRUCKS & CARS IN WORLD WAR

VW AT WAR

KÜBELWAGEN, SCHWIMMWAGEN & SPECIAL VEHICLES

The VW Kübelwagen, Type 82

Michael Sawodny

Jan. 2020

SCHIFFER MILITARY HISTORY
Atglen, Pennsylvania

Sources:
Bundesarchiv, Koblenz
Porsche Archives, Stuttgart
Company Historical and Documentation Office, Volkswagen
AG, Wolfsburg
Scheibert Archives
Podzun-Verlag Archives

I owe special thanks to Herr G. E. J. Kaes, caretaker of the Porsche company archives in Stuttgart who, with great patience and cooperation, allowed me to examine the source material there and provided me with photographs, as well as Dr. B. Wiersch of the VW factory archives, Wolfsburg, and Frau Marianne Loenzartz of the Bundesarchiv Koblenz for their willingness to provide me with information and material. My thanks also go out to Prof. Dr. Wolfgang Sawodny, who generously helped me with this project.

Printed in China.
ISBN: 978-0-88740-308-8

This title was originally published under the title,
VW im Kriege,
by Podzun-Pallas Verlag GmbH, 6360 Friedberg 3 (Dorheim).
ISBN: 3-7909-0108-3.

We are always looking for people to write books on new and related subjects. If you have an idea for a book, please contact us at the address below.

Schiffer Books are available at special discounts for bulk purchases for sales promotions or premiums. Special editions, including personalized covers, corporate imprints, and excerpts can be created in large quantities for special needs. For more information contact the publisher.

Published by Schiffer Publishing Ltd.
4880 Lower Valley Road
Atglen, PA 19310
Phone: (610) 593-1777
FAX: (610) 593-2002
E-mail: Info@schifferbooks.com.
Visit our web site at: www.schifferbooks.com
Please write for a free catalog.
This book may be purchased from the publisher.
Please include $5.00 postage.
Try your bookstore first.

In Europe, Schiffer books are distributed by:
Bushwood Books
6 Marksbury Ave.
Kew Gardens, Surrey TW9 4JF
England
Phone: 44 (0)20 8392-8585
FAX: 44 (0)20 8392-9876
E-mail: info@bushwoodbooks.co.uk
www.bushwoodbooks.co.uk

Professor Ferdinand Porsche with his most famous creation, the Volkswagen, which formed the basis for the military Kübelwagen and Schwimmwagen.

The Volkswagen

Adolf Hitler conceived the idea of the Volkswagen in 1934 soon after he came to power. In his opening speech to the Berlin Auto Show he declared that he considered the design and construction of a "people's automobile" a priority task for the German automobile industry. He later enlarged upon his statement, specifying that the vehicle should offer room for at least four people, consume at most 7 liters of fuel per 100 kilometers, cruise at a speed of 100 kph and cost no more than 1,000 Reichsmarks. Hitler assigned his party comrade Werlin to see to it carried out. Werlin believed that Professor Porsche, who had earlier been active in the design of a small car, and whom Hitler had already enlisted to design a racing-car (Auto-Union-Rekordwagen), was the right man to develop such a vehicle. Financing of the project was to be undertaken by the Reich Association of the German Automobile Industry (RDA), which concluded an appropriate contract with Porsche on 22 June 1934. In the agreement the industrialists set Porsche a very short-term deadline (construction of three prototypes within ten months) in the hope of causing his design to fail, so that they would later be in a position to offer their own "small car" to Hitler as an alternative. Their plan came to nothing, however, on account of Hitler, who valued Porsche highly because of his abilities. He retained Porsche as designer of the car, and in 1935 publicly promised the German people his "people's car," although the first prototypes (V 1 and V 2) were not ready until February 1936, ten months after the deadline set by the RDA.

Three further prototypes (V 3 Series) followed in October of the same year. These

The V 3 prototype (1936) of the legendary VW Beetle, which is still being built today. The basic body shape is already evident, although many details (headlights built into front fenders, doors hinged at front) were later changed.

V 3 *Käfer* (Beetles) already possessed all the characteristic, and at that time revolutionary, features of the VW: the streamlined beetle shape, the torsion bar suspension, patented in 1931, and a four-cylinder, four-stroke, opposed-piston engine which was to remain in use in the VW until the end. The engine had been developed by Porsche engineer Reimspiess following less-than satisfactory trials with two-cylinder four-stroke and two-stroke opposed-piston engines in the V 1 and V 2 prototypes.

Although the extensive tests carried out with these five vehicles had been unequi-vocally successful, the RDA refused to accept the results and demanded a new series of tests with thirty further Volkswagens (W 30 Series). These were built in 1937 and were tested successfully by the "VW Motor Pool" (120 SS men under the leadership of *Hauptsturmführer* Albert Liese), in the course of which the vehicles were driven approximately 2.5-million kilometers. Early in the year Hitler also received a test model of the Volkswagen, with which he was quite taken. Because of the disagreements between the RAD and Porsche, who had American-style (Ford) mass production in mind, Hitler decided to have the Volkswagen produced by a state-owned enterprise and transferred overall control of the project to the party's KdF organization.

The cornerstone for the new VW factory was laid on 26 May 1938 near the town of Fallersleben on the plains of Lower Saxony; the objective was to deliver the first 500 vehicles by the end of 1939. The average citizen was to have the opportunity to purchase the car by paying 5 Reichsmarks per month. However, the outbreak of war halted series production of the VW and deliveries to the purchasers as, understandably, all production was immediately turned to armaments. From 1940 production centred around the military version of the VW (*Kübelwagen*). The 210 "KdF cars" already built were placed at the disposal of senior Nazi functionaries and party officials.

Before the war it was standard practice in the German automotive industry to first design the chassis and then create a body to fit. This made modifications to the superstructure simple and allowed Prof. Porsche to quickly modify the civilian version of the VW into one which could be used by the military.

Here is the steel chassis of the V 3 prototype (the V 1 and V 2 were made of wood) used, not only by the civilian automobile, but, in a slightly modified version, by the Kübelwagen and Type 128 Schwimmwagen as well.

The Kübelwagen

The first mention of a military version of the VW occurred on 11 April 1934 during discussions between representatives of the Reich Chancellery and the Porsche Firm in the Transportation Ministry concerning further development of the civilian Volkswagen. The government representatives wanted the chassis laid out in such a way that it could accommodate three men as well as a light machine-gun and ammunition. Initially, however, this idea was not pursued. The concept was revived by *SS-Hauptsturmführer* Albert Liese, leader of the "VW motor pool" which had carried out road tests with the VW W 30. On 14 January 1938 he made representations to *General der Infanterie* Liese in the *Heereswaffenamt* (Army Ordnance Office, or HWA) and tried to convince him of the VW's potential as a military vehicle. General Liese proved to be interested and that same day directed *Oberstleutnant* Fichtner to draw up an HWA specification for a military version of the Volkswagen based on the drawings provided by *SS-Führer* Liese.

On 26 January 1938 the Wehrmacht specification was presented to Porsche. The vehicle was to be laid out for four soldiers with equipment. It was clear to the Porsche engineers that in order to achieve the specified cross-country capability the vehicle's loaded weight could not exceed 950 kg (150 kg less than the civilian model). As the chassis weighed 400 kg and the planned load capacity had to be 400 kg (100 kg for each soldier and his equipment), only 150 kg was left for the body. In order to achieve this goal entirely new methods would have to be used, as until now the body for a vehicle of this size always weighed about 350 kg. This problem was

solved by Porsche in cooperation with the Trutz firm in Gotha, which had previous experience designing bodies for Wehrmacht vehicles. The HWA also showed interest in a further version with a two-man crew and mounted machine-gun.

On 1 February 1938 the HWA placed an order for the construction of a prototype. It was to be ready for presentation to the HWA on 3 November 1938, allowing nine months for development and construction. Thus the Kübelwagen was born. (The term is an abbreviation of the word "*Kübelsitzwagen*," or "bucket-seat car," referring to the bucket seat standard in Wehrmacht vehicles, not the external shape of the vehicle. The term Kübelsitzwagen therefore applied to any Wehrmacht vehicle and it was not until later that it came to be associated mainly with the successful VW "*Kübel*").

Since the HWA had no objections to the model presented, testing began that same month. Even though it lacked the all-wheel drive considered necessary for an off-road vehicle, the VW Kübel soon proved to be far superior to the standard light car of the Wehrmacht, which had been specially designed to an Army specification. The lack of all-wheel drive was made up for by the vehicle's low weight. The air-cooling of the VW's engine was to be particularly advantageous and proved itself under extreme conditions, be it in the sand of the desert or in the mud and cold of Russia, where the VW Kübel was often the only vehicle capable of functioning. The Wehrmacht's standard light car, on the other hand, failed miserably, often failing to survive 10,000 kilometers of front-line service. Therefore, on 1 November 1941 production was cancelled in favor of the VW Kübelwagen.

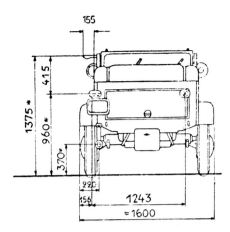

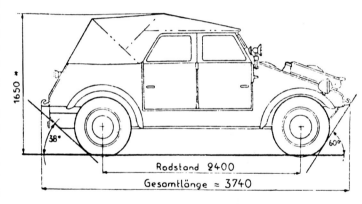

Maßangaben
(* Maße bei Belastung)

The initial version of the VW Kübel (Type 62) was shown to the public for the first time at the 1939 Vienna Automobile Fair. However, the HWA demanded a further improvement in the vehicle's cross-country capabilities. Prof. Porsche achieved this by increasing the vehicle's ground clearance and changing the gear ratio by installing an auxiliary transmission at the rear axle. These changes raised the vehicle's empty weight to

750 kg. The new version was powered by the same 985-cm 3 four-cylinder opposed-piston engine which was also planned for the civilian model. This revised version was designated the Type 82 and entered series production at the VW plant in Wolfsburg in 1940. Bodies for the Type 82 were built by the Ambi-Budd Factory and mounted on the chassis at Wolfsburg.

Among the troops the Type 82 was known and loved everywhere as the Kübelwagen. The 1,000th vehicle left the production line in 1940. A few vehicles saw action in France, but the type first appeared in large numbers during the North African Campaign. The basic Type 82 was delivered to the front in the following versions:
4-seat personnel carrier
4-seat survey vehicle
3-seat light radio car
2-seat casualty evacuation vehicle

From March 1943 the Type 82 Kübelwagen was equipped with the 1,130-cm 3 rebored horizontally-opposed engine used in all-wheel drive types. By the end of the war a total of 55,000 Type 82 vehicles of all versions had been built.

Above right: At the urging of the director of the VW motor pool, *Hauptsturmführer* Liese, the Porsche firm's workshop director, Rudolf Ringel, tried to develop a combat vehicle from the KdF car even before the official specification was issued by the *Heereswaffenamt* (Army Ordnance Office). Seen here is one such early prototype with a mounted machine-gun.

A later prototype of the Geländewagen (Cross-country Car) 62, also called the "Stuka." In contrast to the vehicle illustrated on the facing page, it already displays the angular superstructure with the typical structuring of the sheet metal. The mount for the spare tire (here removed) was located on the side of the vehicle. Note that the windshield is lower on the driver's side than on the passenger's side.

Above: After the Porsche firm was commissioned by the HWA (*Heereswaffenamt*) on 1 February 1938 to develop a military VW, the KdF Cross-Country Vehicle (Porsche Type 62) was conceived on the basis of the VW 38. The prototype was completed on 3 November 1938 and was demonstrated to the HWA. The body did not yet resemble that of the later Kübelwagen, but possessed the rounded shape of the KdF car. The spare tire was stowed on the front hood.

Above right: The same vehicle on one of the test drives which were conducted at the Münsingen Troop Training Grounds from 14 November 1938. Doors were omitted on the cross-country vehicle, being replaced by tarpaulins (rolled up here). The bucket seats are clearly visible.

Right: In order to achieve the cross-country capabilities desired by the Wehrmacht, the Type 62 was developed into the Type 82. Here is a prototype of the latter. In contrast to later production models, the spare tire is recessed in the sloped front hood.

Above and above right: Two factory photographs of a late-production Kübelwagen (1944). Compared to earlier models it exhibits the following changes: the divided rectangular rear window has been replaced by an oval one, and the arrangement of the two exhaust pipes and the rear sheet metal are different.

Right: A view of a production model, revealing the vehicle's interior. Both side doors were attached to a door post as well as hinges.

A view of the engine compartment which housed the four-cylinder, opposed-piston engine, which had a displacement of 985 cm3 and produced 23.5 h.p. at 3,000 rpm (from 1943: 1,130 cm3, 25 h.p.), giving the vehicle a maximum speed of 80 kph. Although the open rear hatch does not provide a view of the flat engine block, various engine assemblies are visible: on the left is the oil-bath air filter with its upward-canted induction pipe, in the center is the Solex downdraft carburettor, below it to the left is the obliquely-mounted distributor cap, and to the right of the carburettor is the generator with its V-belt-driven spur gear. Located at the rear end of the generator shaft beneath the semi-circular sheet metal housing is the fan and next to it the oil cooler, a device which ensured adequate lubrication at high rpm and which previously had only been used in sports cars.

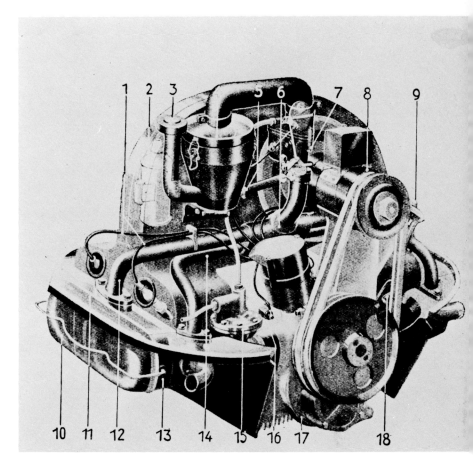

View of the Engine

1. Ignition wires
2. Ignition coil
3. Fan housing
4. Air filter
5. Oil test screw
6. Preheat chamber
7. Carburettor
8. Generator
9. Oil filler nozzle

10. Cylinder cover
11. Gasket
12. Intake Manifold
13. Cylinder Head
14. Preheat Pipe
15. Fuel Pump (mechanical membrane pump)
16. Automatic Switch for oil pressure test light
17. Guard Plate
18. Oil Dipstick

Left: In the summer of 1941 units of the **Afrika Korps** were assembled in Greece, where they carried out training exercises before shipping out for North Africa. The enlisted personnel are seen wearing pith helmets, tropical uniforms and laced boots. The Kübelwagen are armed with an MG 34 machine-gun on a swivel arm.(BA)

Left: Another photograph from Greece taken during a practice drive. In the center is a Mercedes-Benz L 1500 A (Kfz. 18) heavy cross-country truck with a light gun on an improvised trailer.(BA)

Above: An illustration of why the Kübel was so loved by the troops: a light radio vehicle — recognizable by the tall antenna — grinds its way through the desert sand. The VW Kübel saw its first widespread use in the North African theatre — and was a great success.

Above: The VW Kübel in its first appearance in North Africa — during the march-past in Tripoli in February 1941. They are still wearing normal tires, although the crews are ready for the desert war in their pith helmets.

Right: *Feldmarschall* Rommel also used the VW Kübelwagen. He told Porsche that the vehicle had even saved his life when he inadvertently drove into a mine field. Due to its light weight the VW failed to detonate the mines while the heavier Horch car, which was following with the luggage on board, was blown up.

Right: An *Afrika-Korps* column with Kübelwagen (trop). Noteworthy is the Swastika flag mounted on a stepladder at the side of the road, probably as a road marker.(BA)

Above: A brief halt on the North African coastal road before moving on again. The English sign forbids entry into the terrain alongside the road, possibly because of minefield. (BA)

Right: A vehicle the troops were always glad to see: an Army Postal Service Kübel passes a 5 ton half-track vehicle (Büssing-NAG BNT. 8) of the *Afrika Korps* . The half-track is towing an 88mm Flak, which proved so successful against English tanks in North Africa.(BA)

Above: A Kübelwagen drives west along the coastal road. The town might have been Bardia. The primitive road sign, probably erected by the English, displays many of the famous names of the African Campaign. Straight ahead, west of Tobruk, lay Gazala, Acroma and Derna; to the right along the coast, Tobruk; and left, the inland villages of Cubi Hacheim, Capuzzo and Sheferzen, the area of Rommel's disastrous raid of 24/25 November 1941.(BA)

An Army Kübel (trop) transporting Luftwaffe officers in North Africa. The stop was not just to fill the car's front-mounted 40-liter fuel tank, but to have some refreshments as well.(BA)

Right: A Kübelwagen of the DAK (*Deutsches Afrika Korps*) equipped with balloon sand tires in the North African desert. In the background English prisoners are given water.

Above left: A VW 82 (trop) of the Luftwaffe struggles through the softened, rutted terrain of southern Tunisia during Rommel's last offensive at the Kasserine Pass in February 1943. One wonders if the horse shoe on the left fender brought luck — at least enough to allow the driver to survive.(BA)

Above: A photo from the same theatre showing vehicles massed on one of the few, overworked roads. As indicated by the "I" on the side door, this VW Kübel (trop) belonged to a maintenance unit. Serviceable captured vehicles, such as the Jeep, the American equivalent of the Kübelwagen, were taken on strength by German units.(BA)

Left: The same vehicle photographed from behind. It is heading to the rear, while the truck column is taking supplies to the front. The German forces were unable to force the decisive breakthrough. A few days later, after heavy fighting, the Kasserine Pass was recaptured by the American II Corps.(BA)

Right: Speed is of the essence! Pushing a broken-down Kübel under enemy fire. The vehicle probably belonged to *Kampfgruppe Barenthin* which, in November 1942, played a decisive role in frustrating the attempted breakthrough into the rear of the **Afrika Korps** in northern Tunisia following the landings of Allied forces.(BA)

Above: A captured British soldier, apparently wounded in the leg, is taken to the rear in a Luftwaffe Kübel.

Right: This unit of the armored forces in Tunisia was designated *Nahstaffel Sobiak* (note pennant). The vehicle is a Type 82 (trop) equipped with *Kronprinzenrad* (200-12) balloon sand tires.(BA)

Above left: A Luftwaffe unit waits to ship out to North Africa. Behind the VW Kübel in the foreground are two motorcycles, then a Mercedes-Benz Type 170 V which, following the failure of the Standard Light Truck, was an interim solution until sufficient numbers of the VW were available. In line after another VW Kübel are five radio trucks, four of which are mounted on Standard Medium Truck (Kfz. 17) chassis.(BA)

Above: A Kübel drives across one of Tunisia's flat salt lakes at high speed.(BA)

Left: Two Kübelwagen (trop) of *Fallschirmbrigade Ramcke* (note the "R" on the fender). This unit, together with the 164.Inf.Div., was sent to Africa as reinforcements before the Battle of El Alamein. Following the German defeat it withdrew to Tunisia with other units of the **Afrika Korps**, where it was forced to surrender on 6 May 1943.(BA)

A Kübelwagen equipped unit of the motorized Infantry Division (later Panzer-Grenadier Division) *Grossdeutschland* on the march in 1942 (above) and assembling on the broad steppes of Russia (below).(1 x BA)

Right: This striking photograph of VW Kübelwagen in action with an SS Division was published in the *"Illustrierte Beobachter"* in 1941.

17

Above left: The air-cooled engine of the Kübel proved itself during the icy winter of 1941/42 (here in the area of Army Group Center in Russia). When there was a break-down, such as a flat tire, two or three men were sufficient to raise one side of the vehicle off the ground, thanks to its low empty weight of only 775 kg.(BA)

Left: Changing a tire on the same vehicle, under which a jack has now been placed. The photograph provides a good view of the front wheel suspension. On the revolving steering arms, which were mounted on the axle tube, were adjustable mounting bolts which carried the wheel spindle bolts. On these sat the rotating wheel spindle with wheel bearings and brake drum.(BA)

Above: After completing the tire change, the VW Kübel jogs bravely along through the snow-covered Russian countryside. Snow chains have been applied to the rear wheels to improve traction.(BA)

Above: An advance detachment halts before a burning village in the Nevel sector (Army Group North) in December 1942. The two Kübel in the foreground are radio vehicles, as their antennas reveal.(BA)

Right: Even the low weight of the VW Kübelwagen was no help in muddy terrain such as this. The troops at the front will have to wait for the supplies the VW is carrying until the vehicle has been pulled out of this the mud hole.(BA)

Below: *Generaloberst* Lindemann, Commander-in-Chief of the 18th Army in the Leningrad — Volkhov area during a visit to one of the Corps Headquarters under his command. He is using one of the corps' vehicles (note pennant). (BA)

Above: Sometimes even a General needed a push, when the driving rear wheels began to spin in the bottomless Russian mud.(BA)

20

Right: This thin covering of snow (Vitebsk area, early 1944) proved deceptive, as the ground beneath it had already thawed.(BA)

Below: Here a VW Kübel takes on an obstacle more suited to its purpose-built brother, the Type 182 Schwimmwagen (Central Sector, Russia, 1944). The Type 82 had a fording capability of 45 cm.(BA)

Right: Changing the forward axle of a Luftwaffe Kübelwagen. This consisted of two rigid tubes which were fixed to the frame. In these sat the suspension's torsion bars which absorbed shocks transmitted by the revolving steering arms, which are visible on the left. The rubber bumper visible between the steering arms prevented excessive downward movement.(BA)

Dashboard, hand and foot controls.

1. Windshield wipers
2. Horn push button
3. Starter push button
4. Fuse box
5. Socket for portable lamp
6. Switch for dashboard light
7. Battery indicator light
8. Oil pressure warning light
9. Speedometer and odometer
10. Ignition switch
11. Direction indicator light
12. Light switch
13. High beam indicator light
14. Gang switch
15. Fuse box
16. Direction indicator switch
17. Fuel cock
18. Spotlight
19. Foot-operated headlight dimmer switch
20. Clutch pedal
21. Brake pedal
22. Gas pedal
23. Gearshift lever
24. Hand brake lever
25. Choke pull switch

Above left: A Luftwaffe VW on an airfield in France in May 1942. This well-maintained vehicle is wearing hub caps, which were usually dispensed with. In the background is the nose of a Bf 109 G.(BA)

Left: A view of the interior of a Kübelwagen. Notice the scissors-type folding mechanism for the roof, the bucket seats and instrument panel. Mounted on the rear seat hand rail are four brackets for carrying rifles.(BA)

Above: A daylight journey on the roads of France in 1944 often ended like this supply column, which was shot up by Allied fighter-bombers. The crew of the Kübelwagen was lucky, fleeing into cover beneath a tree at the side of the road and thereby escaping unharmed.(BA)

Right: A column of 3 ton half-track vehicles (Sd.Kfz.11) being led in the direction of the front by a Kübelwagen. The guns the half-tracks are towing cannot be seen but, judging by the size of the tire visible behind the first vehicle, they must have been small-caliber weapons. The soldier sitting next to the driver regulated traffic as required with his signal disk. This vehicle, too, carries a horse shoe on the fender as a good-luck charm.(BA)

Facing page: A Luftwaffe unit is transferred from its barracks near Freising to Italy in the winter of 1943. The photographs show vehicles parked in the barracks square, leaving the base and on flat cars during the journey by rail over the Brenner line. Scarcely arrived in Italy and a tire-change becomes necessary. A good view of the raised mounting on the hood for the spare tire, the filler point for the 40-liter fuel tank and the Notek blackout driving light.(4 x BA)

Above right: The Luftwaffe also used the Kübelwagen in the far north, here with *Jagdgeschwader 5 Eismeer* (note unit emblem) in Norway in the summer of 1944. Checking the engine of one of the unit's vehicles.(BA)

Right: A VW Kübel with towing hitch. Here the trailer is loaded with fuel canisters. A Kübelwagen was even tested with a towing hook for a 37mm Pak, however the vehicle (Type 276) did not enter service.(BA)

Special Vehicles

A large number of special variants were based on the basic Kübelwagen design. In 1939 Professor Porsche designed an all-wheel drive version of the VW which could accommodate the civilian body (Type 87) as well as the military one (Type 86). The opposed-piston engine was rebored to 1,130-cm3, which increased its output to 25 h.p. While further development of the Type 86 was dropped in favor of the Type 128 and 166 Schwimmwagen, approximately 600 examples of the Type 87 were manufactured with the KdF car body. These were delivered almost exclusively to the *Afrika Korps*. Like the Kübelwagen, these vehicles were modified for operation in the desert (dust protection for engine, fuel and electrical systems). Following the surrender in Tunisia in May 1943 the remaining Type 87 vehicles were used in the European Theatre.

The hard winter of 1941/42 led the OKW to issue a development contract to the Porsche firm for a vehicle capable of operating in snow. A Type 166 Schwimmwagen (selected because of its flat hull bottom which, it was hoped, would allow it to slide across the snow) was fitted with gripping tires by the firm of Rieger and Dietz. Tests with the vehicle were carried out on the Grosglockner in the period 22-24 June 1942. These revealed that the gripping tires spun easily and tended to bury themselves in snow. An improvement in performance could have been achieved with an increase in the size of the gripping tires and a changed gear ratio; however, a tracked vehicle appeared to offer greater promise. The idea of gripping tires was therefore dropped and the designers turned their attention to the construction of a track-driven type, testing various types of caterpillar drives. Tests carried out in the Black Forest in January 1943 revealed similar problems to the gripping tires: the engine tended to overspeed in the lower gears and the front wheels buried themselves in the snow. Governors were installed in an effort to solve the first problem, while the front tires were equipped with skis, steering being achieved by varying track speed or by differential braking as used in armored vehicles. Although tests were promising, this version of the Kübel did not enter production.

In mid-1943 Senior Government Works Surveyor Dr.Ing. Hanft of the *SS-Führungshauptamt*, together with the Dr. Alpers & Co. Railcar Firm of Hamburg, developed a simple method to allow the Kübelwagen to travel railroad tracks, thereby increasing the vehicle's versatility. The normal wheels were fitted with flanged inserts and the vehicle's track was widened by simple means (achieved on the Type 82 by reversing the wheels) to match the standard railroad gauge (1,435 mm). Porsche created a special reversible drive for the vehicle. Tests were conducted within the Hamburg and Stuttgart State Railway Administrative Areas and were so successful that Porsche was able to demonstrate the Type 82 (schg) to officers of the HWA on 23 September 1943. The officers present were most impressed and agreed to series production, which got under way in the winter of 1943/44. As with many other items of war materiel, however, this was much too late to have any effect on the progress of the war.

These are only three of the more significant special variants based on the VW Kübel-

Above: The Type 82 with the 822 Body as a two-seat siren car. The engine-driven siren was installed in place of the rear seat.

wagen. Additional versions may be seen in the following photographs or are described in the overview on Page 46.

Above: The Type 87, a KdF body on an all-wheel drive chassis. The 564 vehicles of this type which were built were used mostly as staff cars in the North African theatre.

Above right: The rail-capable Kübelwagen Type 157. Its rail travelling capability was achieved simply by installing a flanged insert of pressed steel behind the wheels. The vehicle's track was adjusted to match the standard railway gauge (1,435 mm) through the simple expedient of mounting the wheels in reverse.

The Type 166, which was equipped with Rieger & Dietz gripping wheels, on a test drive in the Großglockner region in June 1942. This idea was soon dropped, however, as the gripping wheels spun easily and tended to bury themselves in the snow.

These three photographs show the different running gear arrangements tested for snow tracks. The drive sprocket was always installed on the rear axle. The track drive extended forward to about mid-vehicle, where it was usually attached to a supplementary swivel arm.

In addition to the version already illustrated, there was also a so-called "long-legged" track drive, in which the track was led back over return rollers. In the photograph below the vehicle is seen climbing a sand dune while on a test drive, where its all-terrain abilities were compared to those of an NSU *Kettenkrad*. The track links have also been fitted with cleats in order to increase traction.

Even a dummy tank (Type 823) was built on a VW Kübel
chassis. It could be used for training purposes, but was
designed chiefly as a means of deceiving the enemy. The panels
over the wheels could be removed so that the vehicle could also
imitate a scout car. The turret rotated, was equipped with
vision slits and could even accommodate a machine-gun. Entry
to the two-seat machine was through a hatch in the turret roof.

The vehicle in the two photographs above is something of a puzzle. The caption accompanying the archive photo states that it was the "Type 166, a shorter, more streamlined four-wheel drive version of the Type 87 developed for the SS." Other sources describe it as a remote-controlled demolition vehicle. The presence of weapons and seats suggest the former, while the lack of instruments tends to support the latter. Noteworthy in any case is the shape of the body which differs sharply from that of the standard Kübelwagen.

Right: In 1941 the *SS-Führungshauptamt* commissioned the Porsche Company with the development of a six-wheeled, cross-country vehicle based on the Type 87. Designated Type 164, the vehicle was to be drivable in both directions and have not only double steering and instrumentation, but two engines as well. The Type 164 progressed no farther than the project stage.

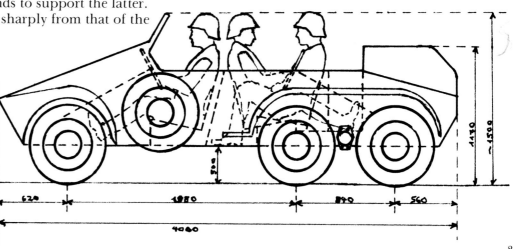

Schwimmwagen

In the mid-1930s the young designer Hannes Trippel was working on the design of an amphibious vehicle, which attracted the attention of the Wehrmacht. In 1939 he was commissioned to design such a vehicle for use by the Army's Pionier units. One thousand examples of the Type 2SG 6, the most successful of the various vehicles created by him, were produced by 1944.

Professor Porsche saw the potential of an amphibious version of the VW Kübelwagen. He later combined these qualities with the all-wheel drive which had been developed at about the same time for the VW (Type 86/87). The result was the Type 128 which appeared in 1940. Thirty examples were built in 1941 at the Wolfsburg Volkswagen works and delivered to the Army's Pionier (Engineer) units. The Type 128 had a boat-shaped body which was installed on the standard Volkswagen chassis. In 1941 Porsche received instructions from the *SS-Führungshauptamt* to further develop the Type 128 which resulted in the Type 166, which later entered large-scale production. The vehicle was to be issued to SS Divisions to replace their motorcycle-sidecar combinations (used by the reconnaissance units), whose off-road capabilities were considered to be inadequate. The Type 166 Schwimmwagen was therefore initially designated a "scout car." The production model, built from 1942, possessed a wheelbase which was 40 cm shorter than the earlier Type 128, while the vehicle's width had been reduced by 10 cm. It was powered by the same 1,130-cm3 engine installed in the Kübelwagen from 1943. In the water the engine drove a three-bladed propeller at the rear of the vehicle. Fully loaded with four soldiers and their equipment, the vehicle's draught was 77 cm. The Schwimmwagen was issued mainly to the divisions of the Waffen-SS but also to Wehrmacht Pionier (Engineer) battalions as well as parachute troops and other elite units. The vehicle was very popular, mainly because of the extraordinary off-road capabilities conferred by the all-wheel drive and its great versatility rather than its amphibious capability, which was seldom used in action. Nevertheless, production was terminated in 1944, because it was believed that further production could not be justified at that stage of the war due to the large number of man-hours and high material usage involved in construction of the vehicle.

Above: Porsche conducted trials to determine the amphibious abilities of the Kübelwagen early in the development program. Here one such sealed vehicle is seen undergoing a test in the Stuttgart firm's fire pond.

Below: This photo of a later Type 128 under test provides a good impression of the front end of the vehicle.

Above: A pre-production model of the Type 128 Schwimmwagen, recognizable by its different engine cover, which also enclosed the propeller, and the special intake and exhaust installations. Here, in contrast to later models, the shaft around which the propeller was lowered extends across the entire width of the vehicle. Below it is the clutch coupling of the intermediate shaft to the motor, which engaged the propeller shaft.

Below: Summer 1941. Two Type 128 Schwimmwagen seen undergoing "swimming" trials in the hands of company employees on the Wörthersee. Prof. Porsche's summer house stood beside this lake.

Above: Wehrmacht officers also tested the Type 128; here one is seen leaving the water after a swimming trip.

Above: One of the 30 Type 128 Schwimmwagen delivered by the VW factory to the Wehrmacht in action.

Below: The advantages of the Schwimmwagen and its four-wheel drive are obvious as it drives through a muddy water hole.

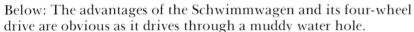

Below: A Type 128 of a Pionier unit takes a slope at high speed. A good view clearly illustrating the shape and ground clearance of the hull.

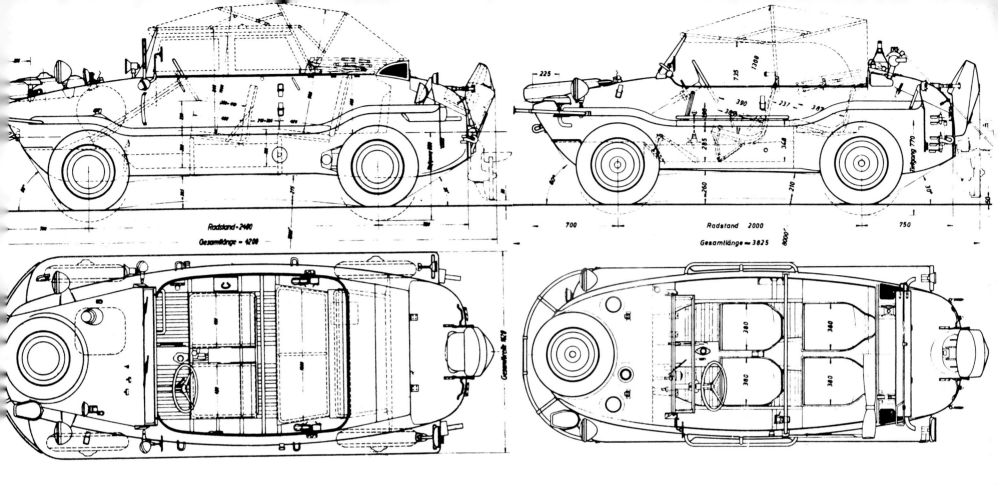

A comparison of these drawings will illustrate the differences between the Types 128 (left) and 166 (right). While the Type 128 had the same chassis as the VW Kübel, the external dimensions of the Type 166 (wheelbase, length and width) were considerably reduced.

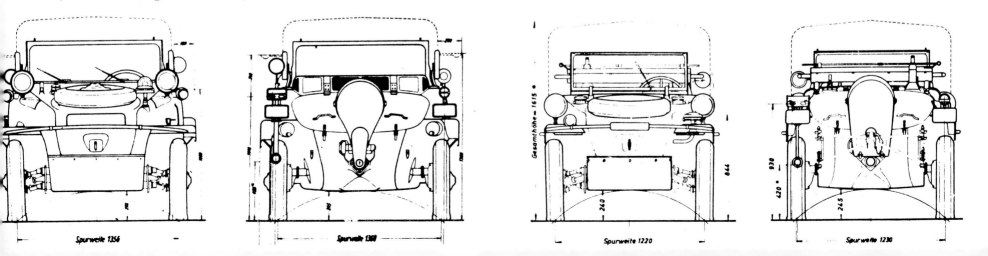

Left: The shorter Type 166 Schwimmwagen which entered large-scale production in 1942. Like the VW Kübel, this vehicle, which was also called a *Kradschützenwagen*, was built in the Volkswagen factories at Wolfsburg.

Right: A rear view of a brand new Type 166 Schwimmwagen. It provides a detailed picture of the engine hood with its hand grips and latches. Above the hood is the exhaust system which, in contrast to that of the Type 128, was located higher and mounted horizontally. On the sides of the metal panel below the exhaust pipes are the air intake vents. Stowed on the exhaust system is the rod used to raise or lower the propeller while the vehicle was in motion. The three-bladed propeller, which is surrounded by a metal shield, rotates around a short shaft. The clutch coupling is clearly visible beneath the propeller, but is covered by a cap attached to the engine cover. On the right and left are swiveling tow hooks.

In 1942 the Schwimmwagen was put to the test by Porsche engineers during an Alpine drive. The machines carried civilian plates during the tests. In the photo above right the first vehicle exhibits several structural differences from later production vehicles, such as mudguards which do not extend the full length of the vehicle, an unbroken top line of the body side panels (as on the Type 128) and the recessed stowage of the spare tire in the front hood. On the other vehicles in the photos the spare tires rest freely on the hood (later production models adopted a middle course: the spare rested freely in front, or was semi-recessed in a bulge in the rear hood). Various pieces of equipment were stowed on the side of the vehicle, such as the shovel and paddle seen on the vehicle on the right. The paddle was carried in case of an engine failure while in the water.

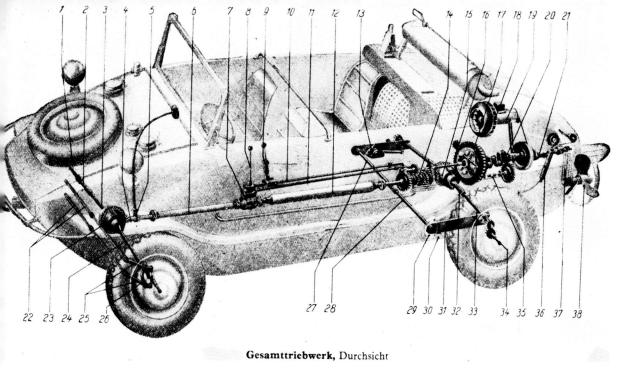

Gesamttriebwerk, Durchsicht

38

Below: Travelling by water in a Schwimmwagen in France. This was probably a test drive, as the vehicle could just as easily have used the road along the adjacent bank. The maximum speed of the Type 166 in the water was 10 kph. The rod for deploying the propeller was fixed to the shield over the exhaust system. (BA)

Above: Layout of the Type 166 Schwimmwagen power train. The crankshaft (20) of the rear-mounted engine drove the rear axle (33) through a gear transmission (14) as well as the front axle (1-3) via a flexible driveshaft (6-12). Forward axle drive and the auxiliary gear with its especially high gear ratio were engaged by a separate lever (9) and shift linkage (10). Coupled directly to the crankshaft behind the generator (19) V-belt pulley wheel was the intermediate shaft for the propeller drive (21).

Right: The vehicle is steered into the water. The *Obergefreiter* mechanic is lowering the propeller with thepurpose-designed rod. The propeller did not have to be secured when the vehicle was travelling in water as the pressure of the water kept it pressed against the clutch coupling. In this way the propeller could fold upwards if it struck an obstacle in shallow water and pass over it. To ensure that no damage resulted, a protective guard was installed in front of the propeller.(BA)

Above: Two officers go for a drive in a Schwimmwagen in a northern French village (summer 1944). As may be seen here, there was also a tow hook on the front of the vehicle.(BA)

Right: The outstanding amphibious capabilities of the Schwimmwagen were much appreciated by members of the Wehrmacht. In this photo the *Hauptmann* (to the driver's right) is about to measure the depth of the water with the bamboo pole in his right hand. The *Obergefreiter* is using the pushrod to hold the propeller in the down position. A rifle rests in the mount on the rear seat handrail. Easily readable is the data plate on the side of the vehicle, which contains the vehicle type (K 2 s), its empty weight (0.83 tons) and pay load (0.45 tons — four soldiers with equipment).(BA)

Left: A Waffen-SS armored artillery battalion is loaded aboard a train in France in early 1944. On the flatcars are "Hummel" self-propelled artillery. In the foreground is a Schwimmwagen with luggage on the rear seat. Hung on the outside of the vehicle in front of the windshield are steel helmets with camouflage covers.(BA)

Below: The crew of a Schwimmwagen observes from a canal bridge in France (August 1944). A tent square is draped over the front hood, while the plants hanging from the vehicle suggest a recent trip through water. In the bracket on the rear seat handrail is a *Panzerfaust* anti-tank weapon.

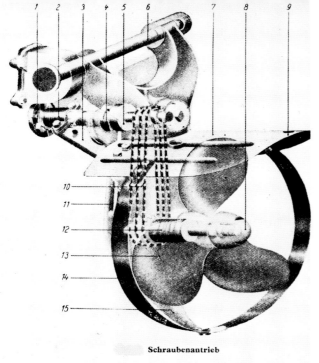

Schraubenantrieb

1 Gewebescheibe	9 Loch für Hubstange
2 Zwischenwelle	10 Ketten
3 Lagerbock	11 Antriebsgehäuse
4 Klaue auf Zwischenwelle	12 Klauenkupplung
5 Klaue auf Antriebswelle	13 Schraube
6 Schwenkwelle	14 Blechkranz
7 Wasserleitblech	15 Bügel
8 Befestigungsmutter für Schraube	

Above: A Schwimmwagen equipped unit in Italy, probably around the time of the Italian cease-fire (September 1943).

Above: Detail drawing of the propeller drive. The intermediate shaft (2) was directly connected to the crankshaft and ended at the rear of the vehicle in a dog (4) into which the upper shaft of the propeller (5) engaged when lowered. The upper shaft in turn drove the lower shaft by a chain drive (10). Because of the direct link between the propeller and the driveshaft, propeller speed could only be regulated by use of the gas pedal and the propeller turned only in one direction (only forward travel was possible).

Right: Checking the rear-mounted engine of a Schwimmwagen belonging to a Panzer unit in Italy. The propeller had to be lowered in order to raise the engine hood. The engine had a capacity of 1,130 cm3 and produced 25 h.p. Power was transmitted to the wheels by a transmission which, in addition to the usual four forward gears and a reverse gear, had a special lever which could be shifted into a supplementary high-ratio gear for cross-country travel, as well as being used to engage front-wheel drive.(BA)

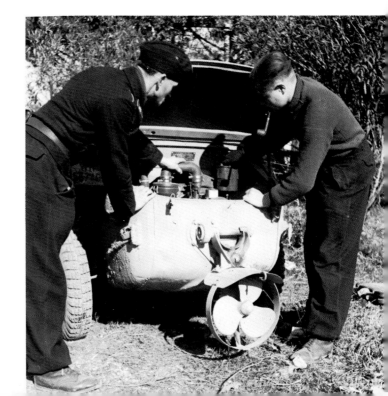

Left: An interesting look into the interior of a Schwimmwagen which is filled with a profusion of items of equipment. Compared to the Kübelwagen, the instrument panel has been significantly revised and simplified. Notice, for example, the smaller and simpler tachometer.(BA)

Facing page:

Above left: The car of the Commander-in-Chief of the 17th Army Headquarters (*General der Infanterie* Allmendinger) with pennant and army insignia on an inspection tour. Below the bar around the front of the vehicle is the shield of the Kuban Bridgehead which, at that time (summer 1943), was being held by the 17th Army. Clearly visible to the left and right of the spare tire are the Type 166's two filler caps for its two fuel tanks which held a total of 50 liters of gasoline (all other vehicles on the VW chassis had one tank of 40 liters). Attached to the step rail on the side of the vehicle (standard equipment) is another step for easier access.(BA)

The next photograph shows the same vehicle from behind, with a clear view of the stowed propeller with protective shield and guard, the clutch coupling, the air intake screen and the exhaust system. The insignia of the army commander is also mounted on the latter.(BA)

Bottom left: This photograph illustrates the fondness of unit leaders for the Schwimmwagen. Here the commander of the Panzer-Grenadier Regiment of the *Grossdeutschland* Division, *Oberst* Lorenz, accompanied by the commander of a Pionier Battalion, carries out a local reconnaissance during the defensive fighting in the Dniepr bend near Kirovograd in the late autumn of 1943.(BA)

Bottom right: This Schwimmwagen, which has just been pushed free after becoming stuck, likewise belonged to the Panzer-Grenadier Division *Grossdeutschland* (note white *Stahlhelm* on rear of vehicle).(BA)

1. Windshield wiper
2. Signal button
3. Fuel cock lever
4. Fuel filter
5. Auxiliary fuel pump
6. Clutch pedal
7. Foot-operated dimmer switch
8. Brake pedal
9. Lever for central lubrication
10. Gas pedal
11. Shift lever for front-wheel drive and auxiliary gear
12. Choke lever
13. Hand brake lever
14. Gearshift lever
15. Ignition switch
16. Speedometer

Above: Even all-wheel drive and amphibious capability are no help here! This Schwimmwagen broke through the ice of a small stream near Nevel in the winter of 1942/43. The crew (in white winter camouflage clothing) have their work cut out for them.(BA)

Above left and below: The Type 166 Schwimmwagen performed its duties reliably in northern Russia, whether during the muddy period (upper photo) or in winter (below). Of special interest in the lower photo are the fuel canisters lashed firmly to the front fenders and the Notek shielded headlight (neither were standard equipment on the Schwimmwagen and were added in the field).(2 x BA)

Once again a Schwimmwagen in action with a Pionier unit.

Above: This is the perspective of the occupants, here while crossing the Beresina River.

Below: Of course the Schwimmwagen could not negotiate a Russian anti-tank ditch. The way had to be shovelled clear.

Overview of Developments Based on the Volkswagen

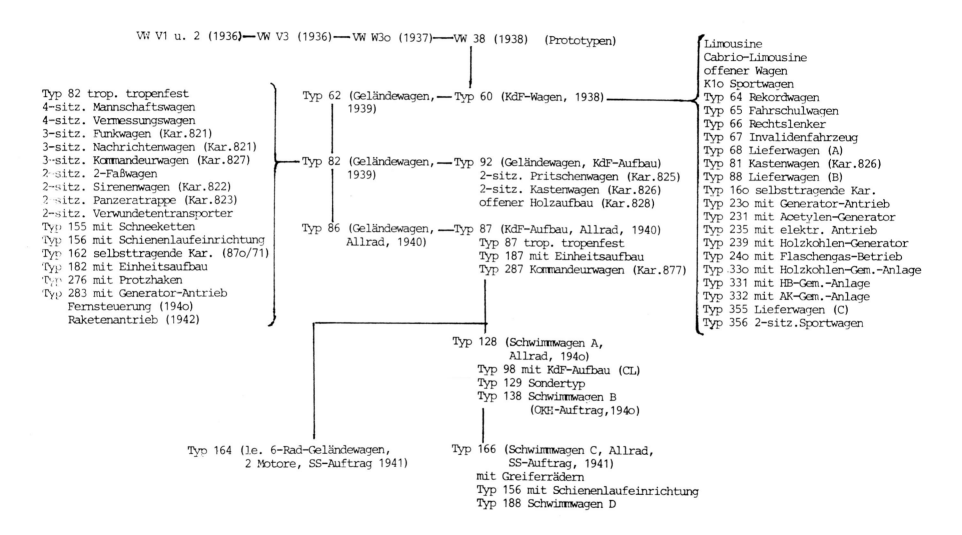

VW V1 u. 2 (1936)——VW V3 (1936)——VW W3o (1937)——VW 38 (1938)   (Prototypen)

Typ 82 trop. tropenfest
4-sitz. Mannschaftswagen
4-sitz. Vermessungswagen
3-sitz. Funkwagen (Kar.821)
3-sitz. Nachrichtenwagen (Kar.821)
3-sitz. Kommandeurwagen (Kar.827)
2-sitz. 2-Faßwagen
2-sitz. Sirenenwagen (Kar.822)
2-sitz. Panzeratrappe (Kar.823)
2-sitz. Verwundetentransporter
Typ 155 mit Schneeketten
Typ 156 mit Schienenlaufeinrichtung
Typ 162 selbsttragende Kar. (87o/71)
Typ 182 mit Einheitsaufbau
Typ 276 mit Protzhaken
Typ 283 mit Generator-Antrieb
    Fernsteuerung (194o)
    Raketenantrieb (1942)

Typ 62 (Geländewagen,——Typ 60 (KdF-Wagen, 1938)
    1939)

Typ 82 (Geländewagen,——Typ 92 (Geländewagen, KdF-Aufbau)
    1939)   2-sitz. Pritschenwagen (Kar.825)
                     2-sitz. Kastenwagen (Kar.826)
                     offener Holzaufbau (Kar.828)

Typ 86 (Geländewagen,——Typ 87 (KdF-Aufbau, Allrad, 1940)
    Allrad, 1940)   Typ 87 trop. tropenfest
                      Typ 187 mit Einheitsaufbau
                      Typ 287 Kommandeurwagen (Kar.877)

Limousine
Cabrio-Limousine
offener Wagen
K1o Sportwagen
Typ 64 Rekordwagen
Typ 65 Fahrschulwagen
Typ 66 Rechtslenker
Typ 67 Invalidenfahrzeug
Typ 68 Lieferwagen (A)
Typ 81 Kastenwagen (Kar.826)
Typ 88 Lieferwagen (B)
Typ 16o selbsttragende Kar.
Typ 23o mit Generator-Antrieb
Typ 231 mit Acetylen-Generator
Typ 235 mit elektr. Antrieb
Typ 239 mit Holzkohlen-Generator
Typ 24o mit Flaschengas-Betrieb
Typ 33o mit Holzkohlen-Gem.-Anlage
Typ 331 mit HB-Gem.-Anlage
Typ 332 mit AK-Gem.-Anlage
Typ 355 Lieferwagen (C)
Typ 356 2-sitz.Sportwagen

Typ 128 (Schwimmwagen A,
    Allrad, 194o)
Typ 98 mit KdF-Aufbau (CL)
Typ 129 Sondertyp
Typ 138 Schwimmwagen B
    (OKH-Auftrag,194o)

Typ 164 (le. 6-Rad-Geländewagen,
    2 Motore, SS-Auftrag 1941)

Typ 166 (Schwimmwagen C, Allrad,
    SS-Auftrag, 1941)
mit Greiferrädern
Typ 156 mit Schienenlaufeinrichtung
Typ 188 Schwimmwagen D

Technical Data:

Engine: (mm)	Stroke	Bore	Capacity	Output	Maximum Torque	Fuel Consumption
	(mm)	(mm)	(cm³)	(h.p./KW)	(mkg at 2,000 rpm)	(l/100 km)
Type 62, 82 (before March 43):	64	70	985	23.5/17.3	6.45	8
Type 82 (after March 43), 87, 128, 166:	64	75	1,130	25/18.4	7.60	8.5

air-cooled, valve-in-head, four-cylinder, four-stroke, opposed-piston engine (compression ratio: 1:5.8, max. rpm 3,300, normal rpm 3,000, with pressure lubrication (oil cooling) and downdraft carburettor

Transmission:
2-wheel drive (Type 62, 82): 4 forward gears, 1 reverse gear (ratios: 3.60/2.07/1.25/0.8 R:6.6)
  with 4-wheel drive (Type 87, 128, 166): an additional cross-country gear (ratio 5.86)
Single-plate dry clutch, stick shift in centre of vehicle

Chassis:
Type 62, 82, 87: torsionally-stiff central canal platform frame with rear bifurcation for acceptance of rear-mounted engine and transmission, mounted sheet-steel body
Type 128, 166: self-supporting, water-tight, sheet-steel hull with double frame side members and longitudinal and lateral bracing
Suspension: Front axle: parallel swinging double-crank axle with single wheel mounting and two transverse torsion bars hydraulic shock absorbers, forward single-acting, rear double-acting front and rear mechanical (cable) 2-shoe brakes, off-road tires 5.00-18 (Type 62), or 5.25-16 (all others), for tropical versions: balloon sand tires 690x200 (200-12)

Dimensions: VW Type	62	82	87	128	166
Length (m)	3.75	3.74	3.85	4.20	3.825
Width (m)	1.55	1.60	1.54	1.62	1.48
Height (m) with roof	1.55	1.65	1.63	1.71	1.615
Wheelbase (m)	2.40	2.40	2.40	2.40	2.00
Track (m)	1.356/1.316	1.356/1.360	1.356/1.360	1.356/1.360	1.220/1.230
Ground clearance (loaded),(m)	0.24	0.275	0.255	0.25	0.24
Turning radius (m)	10	10	10	10	9
Empty weight (kg)	642	668	745	900	890
Loaded weight (kg)	1,100	1,175	1,240	1,398	1,345
Maximum speed (kph)	83	80	80	80	80 (water: 10)
Range (km)	440	440	420	440	520

Above: Naturally, the VW factories were the targets of many bombing raids. This was the Kübelwagen production site in May 1945.

Below: The all-terrain version of the Volkswagen was reborn in modernized form in 1969 as the Type 181, built in the VW plant in Mexico.